WHAT THE FIRST YEAR IN A MARS COLONY WILL BE LIKE
Frontier of Space

True Story of Humanity's First Martian Outpost—Trials, Triumphs, and the Relentless Quest for Survival on the Red Planet

Tommy S. Manley

Copyright ©Tommy S. Manley, 2024.

All rights reserved. No part of this publication may be reproduced, distributed, or transmitted in any form or by any means, including photocopying, recording, or other electronic or mechanical methods, without the prior written permission of the publisher, except in the case of brief quotations embodied in critical reviews and certain other noncommercial uses permitted by copyright law.

Table of Contents

Introduction...3

Chapter 1: Preparing for Mars – The Journey Begins 6

Chapter 2: Building the First Martian Base.............12

Chapter 3: Landing on Mars – A Dangerous First Step...19

Chapter 4: Settling In – The Mars Habitat................26

Chapter 5: Communication with Earth – Staying Connected...33

Chapter 6: Daily Survival Needs – Food, Water, and Air...41

Chapter 7: Science and Research – Pushing the Frontiers of Knowledge... 48

Chapter 8: Health and Well-Being – Maintaining Morale and Health..57

Chapter 9: Challenges of Long-Term Sustainability... 65

Chapter 10: The Role of Robotics – Humans and Machines Working Together......................................73

Chapter 11: Unexpected Challenges and Problem Solving..81

Chapter 12: Looking Forward – Lessons Learned in the First Year...89

Conclusion..98

Introduction

In a world where the limits of human ambition have been stretched beyond Earth's orbit, a new frontier looms large on the horizon. The allure of Mars has captivated humanity for centuries, but what once seemed an impossible fantasy is now unfolding into reality. The journey toward Mars isn't simply about crossing an unimaginable distance in the cold vacuum of space; it's about pushing the boundaries of our endurance, our ingenuity, and our very definition of survival. We're no longer just gazing up at the stars; we're actively carving a path to live among them.

This unprecedented mission to Mars isn't merely an exploration—it's humanity's most daring experiment, one that challenges the spirit of discovery to its core. This mission isn't about touching down and planting a flag; it's about establishing the roots of a living, breathing colony. The ambition is as daunting as it is exhilarating, involving uncharted risks and countless unknowns.

The first year will test everything we know, revealing the unyielding nature of survival on an alien planet. Here, survival takes on a new meaning, where each day's success is hard-won through sheer resilience and breakthrough technologies that bridge the vast gap between dreams and reality.

As you turn these pages, you'll embark on a journey that mirrors the anticipation, challenges, and exhilaration of the first pioneers set to call Mars their home. This book delves deeply into the strategic goals that guide the mission, the towering risks that shadow each step, and the groundbreaking technologies that might just be the lifelines the colony needs to endure. Every innovation and carefully calculated step forward—whether it's generating oxygen from the thin Martian atmosphere or navigating the dangerous landscape with autonomous robots—becomes essential for survival.

Readers will discover the full scope of this high-stakes adventure. Each section brings you closer to the realities that the first Martian settlers will face, from life-or-death struggles to momentous breakthroughs, all forming a mosaic of survival, courage, and relentless human spirit. Expect to explore the daily routines, grueling obstacles, and the extraordinary will required to thrive beyond Earth. As you progress through the book, each chapter reveals a new layer of humanity's interplanetary journey—from the first cautious steps to adapt and survive, to the exhilarating discoveries that will redefine life on Mars and beyond.

For those with a thirst for knowledge about humanity's next great leap, and for anyone who's ever looked up at the night sky and wondered if we're destined to live among the stars, this book offers a front-row seat to the boldest journey humankind has ever undertaken.

Chapter 1: Preparing for Mars – The Journey Begins

The journey to Mars, once a distant fantasy, is now on the verge of becoming reality, thanks to an extraordinary vehicle engineered specifically for interplanetary travel: SpaceX's Starship. This massive, reusable spacecraft stands at the forefront of human space exploration, designed with the singular goal of making Mars accessible. Starship isn't just any spacecraft—it represents years of tireless innovation and a bold vision to enable humanity to reach the stars. Unlike past spacecraft that were built for brief lunar missions or Earth's orbit, Starship is in a league of its own, crafted for long-distance travel, capable of carrying large crews and the cargo necessary for establishing a colony.

Why Starship? To undertake a journey to Mars, the spacecraft must not only be powerful and capable but also built for reusability and sustainability. Starship meets these demands with its next-generation engines, robust cargo space, and its ability to refuel in orbit, dramatically increasing its range. With the capacity to carry about 100 tons of cargo or up to 100 passengers, Starship is designed to make interplanetary travel scalable, something essential for supporting an entire colony on Mars. Other launch vehicles simply can't offer the same payload capabilities combined with full reusability, making Starship the only viable option on the horizon.

SpaceX has outlined ambitious timelines for Starship's missions. Initial projections hinted that the first crewed missions to Mars could be possible by as early as 2026. Yet, space exploration is a complex endeavor, filled with challenges that often demand adjustments to even the best-laid plans. Achieving a Mars landing in 2026 would be a groundbreaking milestone, but as reality often proves, these bold timelines tend to be more aspirational than achievable. A more realistic estimate places the first crewed mission closer to 2030, allowing time for essential testing, repeated simulations, and incremental improvements that ensure Starship can reliably make the journey and safely return.

This 2030 target isn't just about precision or engineering; it's about making sure that every system has been tested rigorously and that the spacecraft can operate flawlessly under the harsh conditions of interplanetary space. By the time the first crew boards Starship for Mars, they'll be embarking on a mission that reflects the culmination of decades of scientific discovery, technological advancements, and human ambition.

The journey to Mars is a test of more than just technology; it's a profound challenge to the human spirit. For the brave individuals chosen to pioneer life on Mars, the physical and psychological demands are intense and unavoidable. Months spent in confined quarters, surrounded by nothing but the deep black of space, can strain even the strongest minds. NASA has long understood these demands, which is why extensive research has gone into studying how space affects human psychology, especially in terms of isolation, confinement, and lack of direct social contact with the world they've left behind. Living in cramped quarters requires each astronaut to adapt to a new definition of personal space, where privacy is limited, and shared living becomes the norm. NASA's research suggests that to preserve mental health, crew members need enough space to both work and unwind. Starship's interior design takes this into account, offering a spacious, pressurized environment with dedicated areas for rest, work, and even recreation.

The psychological challenges, however, extend far beyond physical space. The crew's resilience will be tested by the communication delays with Earth—an unavoidable twenty-minute lag each way that makes real-time conversations impossible. For those on Mars, there will be an inherent sense of isolation, compounded by the realization that even in moments of crisis, immediate help is simply out of reach. To address these risks, NASA's psychologists have developed coping strategies, which include pre-mission training, virtual reality simulations of Earth environments, and consistent mental health support systems. These approaches aim to bolster morale, create a sense of connection, and mitigate the stresses of extreme isolation.

While humans grapple with these psychological and physical challenges, the preparation for a sustainable Mars colony is already underway, with unmanned cargo missions leading the charge. Before any humans set foot on the red planet, a series of autonomous cargo ships will pave the way, delivering essential supplies and equipment necessary for building and sustaining a habitable environment. These cargo missions are designed to carry everything from power generators and habitat structures to rovers and advanced tools. By the time the first crew arrives, these autonomous missions will have delivered everything required to transform the Martian surface into a temporary base of operations.

Autonomous robots are expected to play a crucial role in unloading these supplies and setting up the infrastructure. Equipped with artificial intelligence and precise programming, these robots will handle tasks like assembling shelters, deploying solar arrays, and even preparing landing pads for future missions. This advanced groundwork ensures that when the first human crew finally touches down, they're not arriving to an empty, hostile landscape but to a base that has the essentials for survival already in place. By leveraging this support from autonomous systems, humanity's first footsteps on Mars will be just the beginning of a well-prepared mission to establish life on an entirely new world.

Chapter 2: Building the First Martian Base

Long before humans arrive on Mars, an army of robotic pioneers will take the lead in transforming the barren landscape into a livable base. These machines, equipped with advanced AI and engineering precision, are designed to work autonomously under the harsh Martian conditions, laying the groundwork for the first human settlers. Unlike Earth-based construction, where humans can oversee every step, Mars demands an unprecedented level of robotic independence. These robots will face the relentless Martian dust, extreme temperatures, and the isolation of being millions of miles from human intervention. Their mission is critical: to create a functional, secure environment ready for human life.

One of the most impressive aspects of this mission is the ability of these robots to harness Mars' own resources, minimizing the need to transport bulky building materials from Earth. This concept, known as in-situ resource utilization (ISRU), is at the core of Martian construction. By using what Mars has to offer, from abundant basalt rock to fine regolith (Martian soil), these machines can create a new type of concrete suited to the planet's environment. This "Martian concrete" combines local resources with a biological stabilizer, forming a durable material capable of withstanding the planet's conditions. The structures these robots build will need to withstand both the extreme cold and the constant bombardment of cosmic radiation, making this reinforced concrete an ideal solution.

Robots equipped with advanced 3D printing technology will further transform Martian soil into solid, complex structures, layer by layer. From walls and domes to support beams and landing pads, 3D printing allows for efficient, adaptable construction that can be customized to the specific needs of each structure. These robots will use this technology to construct habitats that are not only protective but also expandable, designed with future additions in mind as the colony grows. Certain designs under consideration include reinforced domes or even cave-like structures, which could provide additional insulation and protection.

As these robotic pioneers work autonomously, they'll also prepare spaces for human activities, from storage areas and laboratories to potential greenhouses. Every structure and system they build will reduce the demands on the first human crew, allowing them to focus on immediate survival rather than base construction. By the time humans arrive, they'll find more than a dusty landing site—they'll be welcomed by a robust, resilient base already equipped for survival. These robotic efforts mark a historic collaboration between human ingenuity and machine capability, setting the stage for a future where humans and robots work together to explore and inhabit the stars.

The first habitats on Mars won't resemble anything from Earth. Instead, these structures are carefully designed to provide protection against Mars' harsh environment. Engineers have drawn up plans for habitats that include artificial caves and egg-shaped towers, each uniquely suited to counter the challenges of radiation, temperature swings, and dust storms that define life on Mars. Unlike traditional buildings, these habitats are focused on function above all else, balancing safety, durability, and efficiency.

Artificial caves are one of the most promising solutions. By digging into the Martian surface, or by constructing cave-like structures with Martian concrete, these habitats gain natural insulation from the surrounding soil, protecting inhabitants from harmful cosmic rays and temperature extremes. Beneath the surface, temperatures are more stable, and the extra layer of protection reduces exposure to radiation—essential for the long-term health of those living in the colony. These artificial caves may also become more than just shelters; they could house essential laboratories, greenhouses, or storage areas, with the insulation making them suitable for storing sensitive equipment or growing food in controlled environments.

Egg-shaped towers are another design under consideration. These tall, rounded structures are intended to minimize exposure to the external environment while maximizing the use of vertical space. Their curved surfaces help distribute stress, making them resilient against pressure fluctuations. Inside, the vertical arrangement allows for a tiered layout, where different levels can serve specific functions—residential spaces, work areas, and relaxation zones, all separated by floors but accessible within one secure structure. These towers, 3D-printed by autonomous robots, would be built layer by layer using Martian concrete and are envisioned as scalable, expandable habitats that can accommodate more settlers as the colony grows.

Tesla Bots, humanoid robots initially developed by Tesla for Earth-based applications, are expected to play a key role in building and maintaining these structures. With SpaceX and Tesla sharing resources and technology, it's highly likely that Elon Musk's humanoid robots will make the journey to Mars alongside other robotic pioneers. These Tesla Bots are designed with versatility in mind, able to perform repetitive, physically demanding, or even delicate tasks that would be challenging or dangerous for humans. From setting up habitat interiors to adjusting equipment or performing maintenance, Tesla Bots can act as both construction assistants and caretakers of the base, freeing up the human crew to focus on scientific research and other complex tasks.

Equipped with a degree of artificial intelligence, these robots can navigate the base, troubleshoot basic issues, and respond to real-time commands from the crew or from mission control back on Earth. The adaptability of Tesla Bots makes them ideal for the unpredictable environment of Mars, where unexpected challenges are inevitable. Over time, they could even be programmed to learn specific tasks as needs arise, evolving alongside the colony and becoming an essential part of daily life. By supporting both construction and maintenance, these humanoid robots add a flexible workforce that complements the capabilities of traditional robots, helping turn Mars from an inhospitable wilderness into a livable home.

Chapter 3: Landing on Mars – A Dangerous First Step

Landing on Mars is a challenge that pushes the limits of aerospace engineering and human ingenuity. Unlike landing on Earth or even the Moon, where real-time adjustments can be made, Mars requires a near-flawless, pre-planned sequence due to the sheer distance involved. Every aspect of a Mars landing must be calculated with precision, accounting for the planet's unique gravitational pull, thin atmosphere, and unpredictable surface conditions. Mars' atmosphere is too thin to provide much aerodynamic braking, yet thick enough that heat shields are essential to avoid burning up during entry. Parachutes, retrorockets, and heat shields all need to work in perfect harmony, executing within seconds of one another to bring the spacecraft safely down to the surface.

One of the most daunting aspects of landing on Mars is that no one on Earth can oversee it in real-time. The vast distance means that it takes approximately 22 minutes for a signal to travel between Mars and Earth. For a landing process that unfolds within minutes, this delay means that any guidance from Earth is impractical. The landing system must be autonomous, capable of making complex decisions in real-time without human intervention. Engineers have equipped Mars landers with highly advanced systems, capable of assessing conditions and making split-second adjustments, but there is always an element of unpredictability.

To address this, spacecraft destined for Mars are tested rigorously, going through simulations that cover countless landing scenarios. Engineers scrutinize every detail, from fuel calculations to the timing of thruster burns, to account for any variable Mars might throw at them. In addition to the technological safeguards, these landers often rely on terrain-relative navigation systems that enable them to identify safe landing zones as they approach the surface. This system uses onboard cameras to compare images of the Martian landscape with pre-loaded maps, adjusting the landing zone to avoid hazardous areas like large rocks or steep cliffs.

The communication delay doesn't just affect the landing process; it also impacts the initial interactions of the crew on Mars. Once on the ground, every message sent to Earth will face this same 22-minute delay each way. For the crew, this means that live conversations are impossible, making Mars feel even more isolated. The need for autonomy is paramount, as the crew can't rely on immediate assistance in the event of an emergency. Every procedure, every system, must be designed to function independently or with minimal guidance.

This delay also impacts the psychology of the crew. Knowing they're physically cut off from Earth reinforces the feeling of being on their own. To ease some of this burden, advanced AI systems on board can assist with diagnostics, maintenance, and troubleshooting, acting as a virtual partner when real-time communication with Earth isn't possible. However, for critical moments like landing, it all comes down to the preparation and reliability of the systems onboard. This leap into autonomy and precision marks a turning point in how we approach space exploration, setting a new standard for missions that venture far beyond Earth's reach.

To ensure a safe landing on Mars, every protocol is meticulously planned with redundancy at its core. Engineers know that even a minor failure could be catastrophic, so they build in multiple safeguards. These safety measures aren't just extras; they're essential backups for each critical phase of descent and landing. Redundant thrusters, backup navigation systems, and additional parachutes are designed to activate if the primary systems experience a malfunction. Each component must perform flawlessly to guarantee that the spacecraft touches down gently and in the correct orientation, reducing the risk to the crew and equipment on board.

Autonomous navigation systems are programmed with failsafes that allow them to adjust in real-time, even without human oversight. Should the main system falter, these backup systems kick in, recalculating descent speeds, adjusting thrusters, or even selecting alternate landing zones if hazards appear unexpectedly. This multilayered approach isn't just about increasing the odds of survival—it's about instilling confidence that, no matter what Mars has in store, there's a plan to handle it.

As the landing sequence ends and the spacecraft touches down, the momentousness of the occasion is overwhelming. The doors open, and the crew steps onto the Martian soil, marking humanity's first physical presence on a world beyond our own. In those initial moments, silence surrounds them, broken only by the faint hum of equipment and the sound of their own breathing within the helmets. Before them stretches a vast, alien landscape, colored in shades of red and ochre, under an open sky tinted by Mars' thin atmosphere. There's an otherworldly quality to the scene, unlike anything they've ever experienced on Earth.

The crew's first steps are cautious, deliberate. They survey the immediate surroundings, ensuring the area is secure and confirming that their equipment has survived the landing intact. Slowly, they begin to move away from the spacecraft, each step imprinted in Martian dust—a historic marker of human perseverance and ingenuity. These first moments are charged with both excitement and gravity. Every movement feels surreal, each glance a confirmation of years of preparation and a symbol of the future possibilities that lie ahead.

This silent Martian landscape is about to become a hub of activity, as the crew begins unpacking supplies, setting up initial systems, and preparing for the long journey of survival and discovery. Each action in these early moments is not just a task; it's the first step in establishing humanity's footprint on

Mars, a testament to the resilience and resolve that brought them here.

Chapter 4: Settling In – The Mars Habitat

Upon arrival, the crew's first shelter will be the Starship itself, repurposed as a temporary home. Compact yet equipped for survival, it offers the basics—sleeping quarters, a galley, and essential workspaces. But as the days on Mars stretch into weeks, the Starship's confined quarters quickly become limiting. Living shoulder-to-shoulder, with only thin walls separating each person's personal space, can place pressure on even the strongest team. Recognizing this, mission planners have prioritized habitat expansion as a critical next step, providing the crew with not only more functional space but also room to unwind and recharge in solitude.

The shift from Starship quarters to a dedicated Martian habitat represents a major milestone in the journey. Soon after landing, the crew will begin work on assembling their expanded living space—using materials brought from Earth in combination with Martian resources. The new habitats, constructed with Martian concrete, will be safer, more spacious, and tailored for the crew's mental well-being. These structures are designed to shield them from cosmic radiation and insulate against temperature extremes while offering more privacy, a crucial factor for maintaining morale over the extended mission.

Each habitat is crafted with attention to personal and communal needs. Private sleeping pods offer individuals their own space to rest and recharge, away from the demands of teamwork. Separate areas are designated for work, exercise, and socializing, creating a sense of routine and balance crucial for mental health. By giving the crew defined spaces for each activity, the design helps prevent the sense of confinement that can arise in an enclosed environment.

As the mission progresses, this expanded space plays an increasingly important role in maintaining mental resilience. The crew is encouraged to create their own routines, allowing moments of solitude and reflection that can make all the difference in such an isolated setting. Whether it's a corner to read, a space for physical exercise, or simply a quiet area to unwind, these expansions provide more than just physical room—they offer relief from the psychological strains of living in close quarters on an alien planet.

This expanded habitat is more than just a shelter; it's a vital structure that supports the crew's mental and emotional well-being. By creating an environment that respects personal space and fosters a sense of community, this carefully designed habitat ensures that the first human colony on Mars not only survives but thrives, turning this isolated outpost into a livable, adaptable, and supportive home away from Earth.

As the crew moves into their newly constructed habitats, the first days are spent testing and adjusting to these Martian structures. Unlike the temporary shelter provided by the Starship, these habitats have been designed to endure Mars' extreme conditions—unforgiving cold, constant cosmic radiation, and the pervasive dust that clings to every surface. The crew must ensure that these new living spaces meet all the vital requirements for a long-term stay. Testing begins with airtight seals, temperature control, and ventilation, each system critical to maintaining a stable, livable atmosphere. They monitor for any fluctuation in pressure or temperature and test air filtration systems to prevent dust accumulation, which could compromise air quality over time.

This period of adjustment is about more than just physical conditions. The crew gradually adapts to the psychological experience of living in an artificial Martian environment. They learn to trust the integrity of their new shelters, where every system has been calibrated to protect them. As each test returns positive results, their confidence in the habitat grows, transforming these structures from cold, technical spaces into genuine homes. With each passing day, they begin to personalize their surroundings, adapting the space to meet their needs and create a semblance of normalcy in this alien setting.

Powering the base is a constant priority, as every system—from life support to heating and lighting—depends on a steady, reliable energy supply. Solar panels, which have already proven their durability on Mars through the long-running rover missions, are the primary energy source. The panels are arranged around the habitat in optimal positions to capture as much sunlight as possible. However, Mars' frequent dust storms pose a challenge. These storms can blanket the solar panels in dust, drastically reducing their efficiency. The crew is prepared for regular dust-cleaning routines, using tools and, when necessary, small robotic assistants to ensure the panels remain clear and operational.

Recognizing the limitations of solar power, mission planners have also considered nuclear power as a backup solution. Compact, self-contained nuclear reactors are among the options that would provide a steady stream of power, unaffected by weather or light conditions. These reactors are designed for safety, with multiple failsafe mechanisms, and could power the colony through the extended Martian nights or during long dust storms when solar power becomes unreliable. While nuclear power is still being carefully evaluated for long-term missions, having it as a backup on Mars could be a game-changer, offering the colony a level of energy independence crucial for their survival.

Between the solar arrays and the potential for nuclear support, the base is equipped with a robust energy plan that underpins every aspect of life on Mars. Powering their new home isn't just about survival—it's a lifeline to exploration, discovery, and growth on this distant planet. With each passing day, the crew solidifies their place in this new environment, inching closer to transforming Mars from a cold, inhospitable landscape into a place where humanity's reach extends just a little further.

Chapter 5: Communication with Earth – Staying Connected

Communication with Earth is essential for Mars missions, not only to maintain a sense of connection with home but also to relay crucial data and updates. However, the challenges of interplanetary communication are vast, and even the simplest exchange can be slowed by the distances involved. SpaceX's Starlink satellite network, initially developed to provide high-speed internet on Earth, holds promise for addressing some of these limitations on Mars. By creating a constellation of satellites around the red planet, SpaceX aims to establish a dedicated Martian communication network, allowing the crew to relay information faster and more reliably than traditional, Earth-based relays.

The goal is ambitious: a network that could eventually transmit data at unprecedented speeds between Earth and Mars, especially during crucial phases like landing and emergency situations.

Starlink's constellation approach, which uses interconnected satellites to relay signals, has the potential to reduce the frustrating communication delays and provide a near-constant link back to Earth. However, this system isn't without its obstacles. One of the biggest challenges is solar interference, which can occur when the Sun positions itself directly between Mars and Earth, blocking signals and creating periods where communication is impossible. Known as solar conjunction, these episodes can last days or even weeks, temporarily isolating Mars from Earth-based support.

To work around these interruptions, engineers are exploring solutions such as relay points positioned in strategic orbits around the Sun. These relay stations could pass messages along without relying on a direct line of sight between the two planets. Additionally, Starlink satellites equipped with optical laser technology are being tested to reduce signal loss by transmitting data through the vacuum of space at the speed of light. This technology, though still in its developmental stages, could transform interplanetary communication by minimizing ground-based relay stations and allowing signals to travel across vast distances with fewer obstacles.

Despite these advancements, Mars' communication system will still require adjustments and backup options to ensure continuous connectivity. During times when solar interference disrupts the connection, the Mars crew must be equipped to operate autonomously, with onboard systems capable of managing daily tasks and emergencies without immediate assistance. Every system on Mars must be resilient enough to function independently, reflecting the reality of this isolated outpost.

Starlink's potential impact on Mars communication is groundbreaking, setting the stage for a new era of interplanetary connectivity. As the technology evolves, the hope is that future Mars colonists will experience a smoother, more direct link to Earth, bridging the vast distance between worlds. Through these innovations, humanity's reach into space becomes more tangible, marking another step toward a future where planets are just points along a universal network, connecting us even as we venture further from home.

To maintain a reliable connection between Mars and Earth, engineers are exploring a range of innovative solutions that stretch the limits of current technology. One promising approach involves establishing relay systems at strategic points along the path between the two planets. These relay stations, positioned in heliocentric orbits, could bypass the issue of direct line-of-sight communication, especially during solar conjunctions, when the Sun's position blocks signals between Mars and Earth. By using satellites stationed at these relay points, the network could route signals around the interference, reducing downtime and offering a more dependable link between worlds.

Laser communication technology is also gaining attention as a possible breakthrough. Unlike radio waves, which are susceptible to interference, lasers transmit data through the vacuum of space with greater speed and accuracy. Starlink's development of optical laser links, which are currently being tested around Earth, could eventually extend to interplanetary distances, allowing information to travel faster with minimal data loss. As this technology matures, it could transform communication between Mars and Earth, making delays less frequent and improving data quality for critical exchanges.

Despite these advancements, there's no perfect solution yet. Communication delays, even under the best conditions, are inevitable due to the sheer distance involved. At their closest, Mars and Earth are still tens of millions of miles apart, and at their farthest, over 200 million miles. Messages sent between the two planets experience a minimum delay of about 20 minutes each way. In practical terms, this means that any instructions or updates from Earth will reach the Mars crew nearly half an hour later, impacting everything from routine operations to crisis management.

The psychological effects of this delay are significant. For the crew, the knowledge that help or advice is never instantaneous amplifies their isolation. Every question, every report, every update becomes a reminder of the distance separating them from home. This delay affects the support structure they're accustomed to on Earth, leaving them to rely heavily on their own judgment and teamwork, fostering a deep sense of self-reliance. To mitigate the psychological impact, mission planners integrate extensive mental health support into training, preparing the crew to manage stress, decision-making, and problem-solving independently.

Operationally, the delay forces the Mars team to adopt a level of autonomy rarely seen in space missions. Systems on Mars must be equipped with advanced AI and automated diagnostics, capable of identifying and addressing problems without waiting for Earth-based guidance. These onboard technologies become the crew's first line of support, designed to handle everything from technical malfunctions to health emergencies. For critical moments, such as landings or surface operations, preprogrammed protocols ensure that the crew can operate smoothly even if Earth's response is delayed.

The combination of relay technologies, laser communication, and autonomous support systems represents humanity's best attempt to overcome the challenges of interplanetary isolation. While these innovations bring Mars closer to Earth in practical terms, the psychological and operational adaptations they require redefine what it means to explore. This new paradigm of self-sufficiency and resilience will serve as the backbone for all future missions beyond Mars, preparing humans for the reality of becoming an interplanetary species.

Chapter 6: Daily Survival Needs – Food, Water, and Air

For the first Martian settlers, sustenance will rely on a delicate balance between shipped supplies and what they can produce themselves. Initially, most of their diet will come from Earth, carefully selected and freeze-dried to retain as many nutrients as possible while maximizing shelf life and minimizing weight. Freeze-dried food offers a familiar, convenient option, providing essential calories, vitamins, and minerals. Yet, as time wears on, there are limitations; certain nutrients degrade over time, and no amount of preparation can recreate the freshness or diversity of a diet that includes real fruits, vegetables, and other perishables. This dependence on pre-packaged food will make fresh, homegrown sources highly desirable for the crew's long-term health.

Growing food on Mars introduces a new set of challenges and possibilities, building on what scientists have learned aboard the International

Space Station through the Veggie Project. This groundbreaking experiment has allowed astronauts to grow plants in microgravity, producing small crops of lettuce, radishes, and other leafy greens in orbit. The ISS experience provides valuable insights into the intricacies of space gardening—like the role of artificial lighting, nutrient delivery, and plant resilience under unusual conditions. Though Mars offers the advantage of gravity, albeit only a third of Earth's, the red planet presents its own unique obstacles, including reduced sunlight, high radiation levels, and a barren atmosphere.

Plants will need more than just seeds and soil to thrive on Mars. Advanced greenhouses equipped with LED lights to simulate Earth-like sunlight and precisely controlled atmospheres to mimic the best growing conditions will be crucial. These systems will allow the crew to experiment with different crops and expand their diet with fresh produce, from nutrient-dense greens to possibly heartier vegetables like carrots or tomatoes. Every fresh food item grown will supplement the freeze-dried rations, not only enhancing nutrition but also providing psychological comfort—a tangible, living reminder of home.

Beyond the nutritional value, space gardening offers psychological benefits. Tending to plants can create a sense of routine and care, giving the crew something to nurture in an otherwise barren landscape. The presence of fresh greens and vegetables will serve as a morale booster, adding color and vitality to the crew's confined habitat.

As the colony grows, so too will their agricultural capabilities. These first attempts at Martian farming will lay the groundwork for increasingly sustainable food production, turning a barren environment into a source of nourishment. While full self-sufficiency remains a distant goal, each successful crop brings the Mars settlers one step closer to independence, where the colony can thrive on resources cultivated from their new world.

Water is one of the most essential resources for life on Mars, but it's also one of the heaviest and most challenging to transport. Initial missions will bring a supply of water, carefully rationed and managed to meet the crew's immediate needs for drinking, hygiene, and small-scale agriculture. Every drop will be recycled to maximize its use, with advanced filtration systems repurposing waste water into clean, drinkable water. But despite these conservation efforts, the supply brought from Earth won't be enough for a long-term settlement, making the search for Martian sources of water a critical priority.

Mars presents a potential solution in the form of its subsurface ice reserves. Certain regions of the planet are believed to hold significant amounts of water ice, buried beneath the surface. The plan is to locate and extract this ice, bringing it into a pressurized habitat where it can be melted into liquid water. However, the Martian ice isn't like Earth's—it may contain contaminants such as heavy metals or salts that render it unsafe to drink without thorough filtration and purification. Technologies developed to desalinate and purify water on Earth will be adapted to handle this extraterrestrial ice, creating a closed-loop system where Martian water, once treated, could sustain the colony indefinitely.

Oxygen, another life-sustaining necessity, poses a similar challenge. Transporting breathable air in adequate quantities isn't feasible, so the solution must come from Mars itself. NASA's MOXIE (Mars Oxygen In-Situ Resource Utilization Experiment) is leading the way in this effort. MOXIE, an experimental device the size of a toaster, has already demonstrated that it can convert Mars' thin, carbon dioxide-rich atmosphere into oxygen. Through electrolysis, MOXIE splits CO_2 molecules, releasing pure oxygen. The device's early tests have been successful, producing enough oxygen to sustain an astronaut for a few minutes. While

MOXIE's output is limited in its current form, the technology shows immense promise for scaling up.

In future iterations, a larger version of MOXIE could be developed to produce oxygen continuously, creating a steady supply of breathable air for a growing colony. This capability would mark a turning point, reducing dependence on Earth and allowing settlers to build an environment that feels a little closer to home. The oxygen produced by MOXIE could also serve other critical functions, from supporting plant growth in greenhouses to providing oxidizers for fuel production, enabling return missions or further exploration of Mars.

Together, the ability to harvest water from Martian ice and generate oxygen from the atmosphere forms the backbone of a sustainable colony. These advances allow the first settlers to adapt their surroundings into a habitable space, transforming Mars from an alien landscape into a place that can support human life. Each drop of water purified and every breath of oxygen created is a step closer to the reality of living on Mars—not as visitors, but as residents of a new world.

Chapter 7: Science and Research – Pushing the Frontiers of Knowledge

The first year on Mars is not only about survival; it's a rare and unprecedented opportunity for scientific discovery. The crew arrives with a clear mission: to study and understand the red planet in ways that remote robots and orbiters never could. With access to Mars' surface, atmosphere, and resources, their research has the potential to unlock answers to long-standing questions about Mars' past, its potential for supporting life, and its suitability for future human habitation.

Geology is one of the highest research priorities. Mars' rocky surface holds clues about the planet's ancient past, possibly revealing whether liquid water ever flowed across its landscape and whether life could have once existed there. The crew will conduct detailed sampling of the soil, rock formations, and minerals, analyzing them in mobile labs or preparing them for return to Earth. These studies could reveal more about Mars' volcanic history, tectonic activity, and the processes that shaped its distinctive valleys and craters. Every sample holds valuable insights, helping scientists compare Mars' geology to Earth's and other celestial bodies, deepening our understanding of planetary evolution.

Atmospheric studies are equally critical. Mars' atmosphere is incredibly thin, composed mostly of carbon dioxide, and provides little protection from cosmic radiation. Studying its composition, behavior, and interactions with the planet's surface and potential weather patterns will provide essential data. Observing seasonal dust storms, temperature fluctuations, and wind patterns is crucial, as these elements directly impact the design and resilience of future habitats and equipment. This research might also yield insights into why Mars lost most of its atmosphere, a mystery that could have implications for understanding atmospheric changes on other planets, including Earth.

Beyond planetary science, the mission is a proving ground for experimental technologies tailored to Mars' unique environment. The crew arrives equipped with new systems and devices, each tested on Earth but never before subjected to Mars' low gravity, extreme temperatures, and abrasive dust. Technologies such as solar-powered energy systems, autonomous robots, and water extraction units are deployed in real conditions, and every piece of equipment undergoes rigorous assessment to gauge its performance, durability, and reliability. The data collected will inform adjustments, helping refine future designs for efficiency and longevity.

One of the most anticipated experimental technologies is advanced robotics. Autonomous and semi-autonomous machines, from rovers to drones, are tasked with exploring the Martian landscape, conducting surveys, and assisting in sample collection. These machines reduce the strain on human astronauts, allowing them to focus on high-priority tasks. Testing the endurance and capability of these robots in Mars' terrain will provide critical insights into their potential role in long-term exploration, setting the stage for more extensive robotic assistance in future missions.

Biomedical research is another key component of the crew's scientific goals. Mars' low gravity and high radiation levels pose unknown risks to human health, so monitoring the crew's physical and mental well-being is essential. This data not only helps protect the current team but also builds a foundation for understanding how the Martian environment affects the human body. Everything from muscle and bone density changes to psychological resilience will be documented, offering valuable insights for future settlers and explorers.

These research and testing efforts form a foundational layer for understanding Mars and preparing for the future. Every sample collected, every piece of data analyzed, and every experimental device evaluated brings humanity closer to transforming Mars from a scientific curiosity into a new frontier for human life. The discoveries made in this first year may shape how humans approach exploration, colonization, and even survival beyond Earth, inspiring a new era of scientific and technological innovation driven by the challenges of an alien world.

The first settlers on Mars are not just explorers; they are also the subjects of one of the most ambitious medical and biological studies in history. Living on Mars means confronting an environment unlike any humans have ever endured, with gravity only a third of Earth's, constant exposure to cosmic radiation, and the physical and psychological demands of isolation. For researchers on Earth, the effects of these factors on human health are still largely unknown. Monitoring the crew's biological responses to their surroundings is crucial for understanding the full impact of this new habitat.

Low gravity presents a unique set of physiological challenges. On Earth, gravity shapes everything from bone density to muscle strength, and our bodies are optimized for life under its pull. In a reduced gravity environment like Mars, the body begins to undergo changes that could lead to muscle atrophy, reduced bone density, and shifts in fluid distribution. Daily exercise routines tailored to combat these effects are essential, and the crew will perform regular strength and endurance tests, providing data that informs future strategies to maintain health in low-gravity environments. Every aspect of their physical adaptation is documented, with measurements, scans, and biometric data collected to help scientists design more effective countermeasures for future missions.

The Martian atmosphere offers little protection against cosmic radiation, which poses a long-term threat to human health. Despite shielding in the habitat and protective suits, the crew is exposed to levels of radiation far beyond what they would experience on Earth. Radiation exposure can damage cells and increase cancer risk over time, making it one of the most significant concerns for long-duration missions. Monitoring radiation exposure is essential, and the crew carries personal dosimeters to measure accumulated levels. Through this constant tracking, scientists can assess how well the current protective measures work and explore potential ways to enhance them, from improved habitat shielding to new materials for spacesuits.

Psychological resilience is just as crucial. The isolation of Mars, combined with the communication delay to Earth, requires the crew to be more self-sufficient than any astronauts before them. They rely on one another for social interaction, support, and problem-solving, as there is no real-time connection to mission control. The crew's mental health is closely observed, with regular assessments to gauge mood, stress levels, and team dynamics. Tools like virtual reality are used to create immersive environments that help mitigate the effects of isolation, allowing the crew to experience familiar scenes from Earth. This psychological data is invaluable, guiding future missions on how best to support crews on long-term, isolated expeditions.

Beyond the walls of their habitat, Mars itself awaits exploration. The initial rover missions are essential for mapping the terrain surrounding the base, providing the crew with an understanding of their immediate environment. Equipped with advanced sensors and cameras, the rovers autonomously navigate the Martian landscape, identifying geological features, potential hazards, and areas of interest. This mapping not only informs safe paths for the crew's excursions but also guides scientific research, pinpointing locations where the soil, rock formations, or other features warrant closer study.

These rovers act as the crew's scouts, extending their reach and reducing the risks of navigating the unknown. In addition to terrain mapping, they conduct basic scientific surveys, examining the chemical composition of rocks and searching for any signs of past or present water activity. Each rover's findings are relayed back to the habitat, allowing the crew to plan excursions with precision and purpose.

Together, the medical and biological research within the habitat and the exploratory work of the rovers beyond it create a comprehensive picture of life on Mars. The data gathered during this first year will inform nearly every aspect of future missions, from health protocols and habitat designs to safety procedures and exploration tactics. For now, these pioneers are adapting and exploring, each day bringing them closer to understanding the red planet—and preparing humanity for a future where living on Mars is no longer a distant dream, but a sustainable reality.

Chapter 8: Health and Well-Being – Maintaining Morale and Health

Life on Mars challenges the human body in ways that Earth-bound physiology simply isn't prepared for, especially under Mars' lower gravity. Without Earth's constant pull, muscles and bones can begin to weaken, leading to a gradual loss of strength and endurance. To combat these effects, the Mars crew adheres to a strict fitness regimen designed specifically for low-gravity environments. Equipped with specialized exercise machines tailored for resistance training, each astronaut engages in daily workouts that target muscle groups most susceptible to atrophy. Resistance bands, compact weight systems, and treadmill harnesses simulate the load-bearing movements familiar on Earth, helping to maintain muscle tone and joint health. These exercises are essential for more than just physical fitness—they are critical for the long-term health and mobility of each crew member, ensuring they can handle both the demands of Mars and the eventual return to Earth's gravity.

Maintaining bone density is equally important. The lower gravitational pull on Mars causes calcium loss in bones, which can lead to osteoporosis if left unchecked. High-impact, weight-bearing activities are incorporated into the crew's routine as much as possible, simulating the forces that would naturally strengthen bones on Earth. Each exercise session is monitored carefully, with data sent back to Earth for analysis. These insights not only help scientists improve the crew's regimen but also pave the way for future exercise strategies that can support health in other low-gravity environments, like the Moon or even deep space.

While physical fitness keeps their bodies strong, the crew's mental health is safeguarded with equal rigor. Mars is not just physically isolated from Earth; it's also profoundly remote on a psychological level, with a silence and solitude unlike anything found on Earth. Long periods without direct human contact can lead to feelings of loneliness, anxiety, and even depression. To counteract these effects, mission planners have equipped the crew with virtual reality tools that recreate familiar, comforting environments from Earth. With a VR headset, an astronaut can escape to a virtual beach, walk through a forest, or even simulate visits with loved ones. These immersive experiences provide a mental respite, offering a sense of familiarity in an alien landscape and helping crew members manage stress.

In addition to virtual reality, the crew relies on structured social routines, regular team-building activities, and dedicated relaxation time. Scheduled calls with mental health professionals on Earth provide additional support, allowing crew members to discuss their experiences, frustrations, and triumphs with someone outside the immediate team. The crew also receives support through specially tailored programs that encourage self-reflection, mindfulness, and resilience. These mental health strategies are designed to create a sense of stability and grounding, even when Earth feels worlds away.

Together, these physical and mental health practices are as essential to survival as the Martian habitat itself. By focusing on both body and mind, the mission not only enhances the crew's quality of life but also ensures they can meet the extraordinary demands of life on Mars with strength and resilience. The knowledge gained from these practices will guide future interplanetary missions, where the importance of holistic health care becomes paramount in enabling humans to live and thrive in the vast unknown.

Living and working on Mars requires a level of teamwork that few on Earth will ever experience. In a small, isolated group, every interaction, every decision, and every shared task affects the team's overall cohesion. The challenges of life on Mars—constant confinement, lack of privacy, and high-stakes problem-solving—create a unique pressure cooker where minor disagreements can escalate if not managed carefully. Here, there's no option for a break from the group, no escape from the habitat walls, and no immediate help if conflicts arise. To prepare for these social dynamics, each crew member undergoes rigorous training before departure, focusing on communication, conflict resolution, and adaptability. They learn to recognize their own stress responses and those of others, fostering a sense of empathy and understanding that becomes vital in an environment where mutual reliance is key.

Structured routines also play a crucial role in maintaining team harmony. With daily schedules that balance individual tasks and group responsibilities, the crew develops a rhythm that reduces friction and creates stability. Regular debriefs, where each member can share concerns or suggest adjustments, help maintain transparency and allow any brewing tensions to be addressed early. Social activities are encouraged, even within the confines of the Martian habitat. Movie nights, team meals, and recreational hours become valuable moments of bonding, reminding the crew that they are not just colleagues but a support system for one another.

In an environment where mistakes can have dire consequences, emergency preparedness is paramount. Mars leaves no room for error, and the crew must be ready to handle both physical and psychological emergencies with the limited resources available. Each crew member is trained in basic medical procedures, equipping them to manage injuries, illnesses, and even mental health crises in the absence of an Earth-based medical team. The habitat is stocked with essential medical supplies, and every procedure is carefully mapped out, from wound care to setting fractures, ensuring that all crew members are capable of responding swiftly to unexpected situations.

For more severe cases, telemedicine provides a lifeline. Through video links, the crew can consult with doctors and specialists back on Earth, receiving guidance for advanced medical issues. However, the time delay requires them to be self-sufficient, able to perform treatments and make decisions under pressure. Psychological emergencies are handled with equal diligence. Each member has access to mental health support, including counseling sessions over video link and tools for self-care and stress management. Protocols are in place to recognize early signs of psychological strain, allowing the crew to provide support for each other or seek remote intervention if necessary.

Emergency preparedness on Mars is as much about mindset as it is about skills. The crew's training instills a sense of calm in the face of crisis, reminding them that they are equipped to handle whatever comes their way. Knowing they have the resources and support systems in place to address both physical and psychological challenges strengthens their resilience, empowering them to face the uncertainties of life on Mars. This proactive approach to social dynamics and emergency readiness ensures that, whatever the red planet throws their way, the crew can stand together as a unified, capable team.

Chapter 9: Challenges of Long-Term Sustainability

As the Martian colony matures, the goal shifts from survival to sustainability, and nowhere is this more critical than in the development of a self-sustaining food system. Initially, the crew depends heavily on freeze-dried food supplies brought from Earth, but over time, they will work towards cultivating a reliable source of fresh produce, ensuring that the colony can reduce its dependency on Earth shipments. The colony's greenhouses, designed to replicate the essential conditions for plant growth, will host crops carefully chosen for their nutritional value, fast growth cycles, and low resource requirements. Leafy greens, root vegetables, and grains top the list, providing essential vitamins, minerals, and calories to complement their diet.

The journey toward food independence also opens the door to discussions around protein sources. Although raising traditional livestock on Mars is impractical due to space and resource limitations, the crew can explore alternative protein options suited to their environment. Insects, such as crickets or mealworms, offer a promising solution, as they require minimal water, space, and feed to thrive. These high-protein sources are sustainable, nutritious, and efficient, providing a versatile ingredient that can be added to various dishes. Another potential breakthrough lies in cellular agriculture—growing lab-based animal proteins in controlled environments. While this technology is still developing on Earth, the unique demands of Mars could accelerate its progress, potentially allowing the crew to produce meat-like protein with minimal resources.

Equally essential to sustaining life on Mars is water—a resource that must be managed with extreme care. The initial water supply, transported from Earth, is precious, but it is far from infinite. As such, water recycling and conservation systems are integral to the colony's long-term survival. Advanced filtration units enable the recycling of almost every drop, turning wastewater from showers, washing, and even bodily fluids into drinkable water. This closed-loop system is based on techniques pioneered on the International Space Station, refined further to handle the greater volume required for a Martian outpost. Each liter of water is filtered, purified, and reused countless times, allowing the crew to operate with remarkable efficiency.

Water conservation practices go hand in hand with recycling, as the crew learns to maximize every use before reclaiming it. Plants grown in the greenhouse are carefully watered to prevent wastage, often through drip irrigation or hydroponic systems that deliver water directly to the roots with minimal loss. Humidity from the habitat's air is also captured and condensed, converting atmospheric moisture into usable water.

In time, the crew may also tap into local water sources, such as subsurface ice deposits, supplementing their recycled supply with Martian water. This ice, once processed and purified, can integrate seamlessly into the colony's water system, further enhancing their self-sufficiency. Each step taken toward water independence reduces the need for shipments from Earth, moving the colony closer to a sustainable existence.

Together, the strides made in food and water independence mark the foundation of a self-sustaining life on Mars. With each success, the colony inches closer to a reality where it can thrive on what Mars has to offer, transforming an alien environment into a home where human life can flourish autonomously.

Maintaining a stable supply of breathable air is one of the core challenges for a Martian colony, and at the heart of this endeavor is NASA's MOXIE technology. Originally developed as a compact prototype, MOXIE (Mars Oxygen In-Situ Resource Utilization Experiment) has demonstrated that it can convert the carbon dioxide-rich Martian atmosphere into oxygen. While the device's current output is modest—producing enough oxygen to sustain an astronaut for a brief period—it represents a crucial first step. For the colony to truly become self-sufficient, this technology must be scaled up significantly, with larger units capable of generating oxygen continuously to meet the needs of multiple inhabitants.

Scaling up MOXIE involves more than just increasing its size. Engineers must enhance its efficiency, durability, and integration with the colony's habitat systems. Larger, upgraded MOXIE units will likely be embedded into the habitat infrastructure, running around the clock to maintain an ample oxygen supply for breathing, as well as other essential processes like greenhouse support and potential fuel production. As the colony grows, so will the demand for oxygen, making this technology indispensable for the vision of a sustainable Martian settlement. The goal is to reach a point where oxygen production on Mars becomes routine, removing the need for oxygen shipments from Earth and enabling the colony to operate autonomously.

Achieving this level of self-sufficiency with air quality is part of a larger transition that the colony will undergo as it moves away from dependence on Earth-supplied resources. At the outset, Earth provides the colony with essentials, from food and water to construction materials and medical supplies. Yet, every shipment from Earth is costly, limited by launch windows, and subject to delays. As such, in-situ resource utilization (ISRU)—the practice of using local materials to meet the colony's needs—is the cornerstone of Mars colonization.

The transition toward autonomy involves harnessing Mars' natural resources to produce essentials locally. Water extraction from subsurface ice deposits, nutrient recycling for agriculture, and the development of Martian concrete from regolith are just a few examples of ISRU in action. Each in-situ solution decreases the colony's dependency on Earth, allowing settlers to live off the land, so to speak, and fostering a sense of resilience and adaptability. With each system that shifts to local sourcing, the colony moves closer to becoming a self-contained ecosystem capable of sustaining itself indefinitely.

This shift isn't just about survival—it's about establishing a foothold that can expand and evolve. A colony capable of producing its own food, water, oxygen, and building materials opens the door to growth. Over time, new habitats can be constructed using local resources, and the infrastructure can expand to support more settlers, larger research facilities, and a richer quality of life. In this way, autonomy isn't simply a logistical goal; it's the key to transforming Mars from an outpost into a permanent human settlement, a place where future generations might live, work, and call home.

Each step toward independence reinforces the colony's viability, proving that life on Mars can be more than a temporary experiment. By mastering air, water, food, and construction from local resources, humanity lays the groundwork for a new chapter in its journey, creating a sustainable existence on a world far from Earth but increasingly self-sufficient in its own right.

Chapter 10: The Role of Robotics – Humans and Machines Working Together

In the ambitious quest to build a sustainable colony on Mars, humanoid robots and other autonomous machines are invaluable allies. Tesla Bots, designed with human-like dexterity and precision, are among the key players in this endeavor, capable of performing physical tasks that would be demanding, time-consuming, or even dangerous for human settlers. With adaptability built into their programming, these humanoid robots can handle various jobs, from assembling habitats and unloading cargo to maintaining equipment and performing repairs. Tesla Bots are specifically suited to environments like Mars, where reliable assistance is essential, and each robot's ability to adapt to different tasks brings flexibility to an otherwise rigid mission structure.

The robots' resemblance to human movement and function enables them to perform intricate tasks that require a fine touch—adjusting machinery, setting up solar panels, and even handling basic repairs to habitats or equipment. By working alongside humans, these robots not only extend the crew's capabilities but also allow the team to focus on more complex scientific and operational challenges. They act as an extension of human effort, mirroring human motions and carrying out commands with remarkable precision, reducing physical strain on the crew and accelerating the construction of Mars' infrastructure.

Meanwhile, autonomous rovers and drones expand the reach of exploration far beyond the immediate base. Equipped with advanced sensors, mapping tools, and cameras, these machines navigate Mars' rugged terrain independently, covering vast areas that would be impossible for humans to reach on foot. Rovers, designed to withstand Mars' dust storms and temperature extremes, can spend days or even weeks in the field, conducting geological surveys, collecting soil samples, and analyzing mineral compositions. Each rover's findings contribute to a growing map of the Martian landscape, identifying areas of scientific interest and potential resources, from subsurface ice deposits to rich mineral zones.

Drones, equipped to handle Mars' thin atmosphere, provide a different perspective altogether. By flying above the surface, they offer a bird's-eye view of the terrain, capturing high-resolution images and identifying features that might otherwise go unnoticed. These drones are especially useful for scouting new paths, mapping hazardous areas, and monitoring weather patterns from above. Their speed and agility make them ideal for rapid surveys, while their ability to recharge at the base allows for continuous exploration.

Together, these robots—humanoid, rover, and drone—create a dynamic, multi-layered exploration system. While Tesla Bots focus on establishing and maintaining the colony itself, rovers and drones push outward, gathering crucial data about Mars' geography, climate, and resources. The integration of these technologies represents a symbiotic relationship between human and machine, where each complements the other's strengths.

This network of robotic assistance not only accelerates the pace of development on Mars but also mitigates risk by performing high-exposure tasks autonomously. By mapping the surrounding landscape and constructing essential infrastructure, these machines build the foundation for future expansion, ensuring that the Mars colony grows from a small outpost into a thriving community. Each task they accomplish, from placing a structural beam to surveying distant terrain, brings humanity closer to transforming Mars into a world where we can live, explore, and thrive.

Machine learning and artificial intelligence play a pivotal role in adapting technology to the unique challenges of Mars, where conditions demand flexibility, rapid adaptation, and autonomy. The red planet's unpredictable environment, from intense dust storms to extreme temperatures, necessitates that machines be able to learn, adjust, and optimize their performance on the fly. Through machine learning algorithms, rovers, drones, and Tesla Bots can continually refine their actions, evolving from simple pre-programmed tasks to more complex operations based on real-time experiences. AI allows them to recognize patterns, such as changes in terrain or early indicators of mechanical wear, and respond accordingly, reducing the likelihood of breakdowns and enhancing efficiency.

These technologies are also crucial for energy management. On Mars, every watt of power counts, so machines are programmed to adjust their energy consumption based on tasks and environmental conditions. For instance, a rover may enter a low-power mode during dust storms when solar energy is limited, preserving its battery until the skies clear. By using predictive algorithms, these machines can make real-time decisions about when to conserve power, when to deploy sensors, and how to prioritize tasks based on the conditions at hand.

Yet, despite the impressive autonomy that machine learning and AI enable, human oversight remains essential, striking a balance that ensures the machines align with mission objectives. While AI can recognize patterns and learn from previous tasks, there are still unforeseen scenarios where human judgment and creativity are indispensable. Crew members can provide nuanced decisions, course corrections, and new objectives that are beyond current AI capabilities, especially when encountering unexpected obstacles or prioritizing high-stakes tasks. Mars is an environment where creativity, intuition, and adaptability are essential for survival—qualities that, for now, remain human strengths.

The collaboration between humans and robots on Mars is both delicate and interdependent. Humans rely on robots to extend their reach, handle strenuous tasks, and minimize exposure to environmental hazards. In return, robots rely on humans for mission direction, problem-solving insights, and regular maintenance. This symbiosis requires trust in the reliability of machines and the flexibility to intervene when necessary. For instance, while Tesla Bots are designed to operate with minimal oversight, they still depend on human team members to troubleshoot complex malfunctions, reboot systems, or recalibrate sensors in ways that AI cannot yet replicate.

In practice, this partnership allows for a layered approach to task management. Robots undertake routine, repetitive, or hazardous tasks, while humans focus on high-priority objectives, like scientific research, strategic planning, and complex troubleshooting. By sharing responsibilities, human-robot collaboration maximizes productivity and ensures the colony operates smoothly, even in a challenging, isolated environment. Over time, this balance evolves, with humans gradually teaching the machines through adaptive algorithms, fine-tuning the AI's responses, and shaping its decision-making processes based on practical experiences.

The careful balance between automation and human intervention on Mars paves the way for future missions to venture even deeper into space. Each lesson learned from this partnership refines AI's role in exploration, fostering a world where humans and intelligent machines coexist in distant frontiers, pushing the boundaries of what's possible and expanding humanity's footprint across the cosmos.

Chapter 11: Unexpected Challenges and Problem Solving

Mars is a planet of extremes, and its environment brings with it an array of surprises that are difficult to fully predict or prepare for. Dust storms, one of Mars' most iconic features, are more than just a visual spectacle—they're a relentless challenge for equipment and daily operations. These storms can blanket the entire planet, with winds stirring up fine particles that cling to solar panels, clog filters, and interfere with machinery. Although Martian storms don't have the force to knock over structures, the fine dust reduces visibility, coats essential equipment, and blocks sunlight, cutting off the solar energy that powers much of the base. For the crew, preparing for these storms means carefully managing energy reserves, rationing power, and devising methods to clean and maintain solar arrays once the skies clear.

Mars' temperature swings are another constant challenge, with daytime highs above freezing and nighttime lows plunging to depths that can strain even the most resilient machinery. The thermal stress can cause materials to expand and contract rapidly, leading to cracks, warping, and fatigue in metal and plastic parts. Electronics, too, are vulnerable to these extremes, and without the natural buffering that Earth's atmosphere provides, the crew must rely on insulated habitats and equipment storage to minimize damage. In this volatile climate, every piece of technology is pushed to its limits, testing the durability of materials and the foresight of mission planners.

Yet, even the most carefully designed equipment will encounter issues on Mars, and malfunctions are an inevitable part of daily life. With resupply missions few and far between, the crew has no option but to address these issues with limited tools and materials on hand. Repairs must be approached with creativity and adaptability, and every piece of equipment is evaluated not just for its primary use but for how it might be repurposed or repaired if something breaks. Duct tape, 3D-printed components, and spare parts salvaged from non-essential items become lifelines, transforming the colony into a workshop of innovative problem-solving.

For critical repairs, the crew relies on a combination of resourcefulness and technical training. Each member is cross-trained in various maintenance skills, from basic electronics repair to mechanical troubleshooting, allowing them to step in whenever a piece of equipment falters. In situations where complex machinery malfunctions, onboard diagnostic tools and remote guidance from Earth can assist, though the communication delay requires them to make quick, informed decisions without immediate feedback. Every repair task reinforces their adaptability and resilience, turning each setback into a learning opportunity that prepares them for future challenges.

In the face of unanticipated obstacles, the crew's ability to improvise and endure becomes one of their greatest assets. Environmental surprises test the limits of their equipment, but they also highlight the colony's resourcefulness and unity. Each storm weathered and each repair completed brings the team closer to mastering life on Mars, preparing them for a future where they can confront the unexpected with confidence, turning Mars from an alien landscape into a home where they can not only survive but thrive.

On Mars, health crises are not just challenges—they are high-stakes events that test the colony's preparedness and resilience. Far from Earth, where advanced medical facilities and specialists are a quick call away, the crew faces the reality of having to manage most medical emergencies on their own. Every crew member undergoes extensive medical training before leaving Earth, learning essential skills for handling injuries, sudden illnesses, and even surgeries with minimal equipment. The Martian habitat is outfitted with a compact but well-equipped medical bay stocked with essential supplies and diagnostic tools, from basic wound care kits to more advanced imaging and surgical instruments.

In the event of a medical crisis, crew members must act quickly and decisively, relying on protocols carefully developed to address various scenarios. For more routine issues like sprains, minor cuts, or infections, the crew has the resources to provide effective care. However, serious injuries or illnesses demand an organized response. Telemedicine provides an additional layer of support, enabling crew members to consult with Earth-based doctors, though the communication delay necessitates a level of self-reliance. For critical situations, remote medical experts can guide the crew through complex procedures step-by-step, yet the responsibility ultimately rests with those on Mars. Preparedness and calm under pressure are vital, as every second counts in an environment where even a small medical setback can escalate.

Beyond the physical risks, the psychological strain of life on Mars is profound. The isolation, confined quarters, and distance from Earth create an environment where emotional resilience is as essential as physical health. Real-life scenarios could include periods of heightened tension, interpersonal conflicts, or moments of acute homesickness, all of which can affect individual morale and group dynamics. In these moments, the crew relies on mental health strategies they have been trained to use—mindfulness exercises, regular physical activity, and structured social interactions. Scheduled downtime and private spaces within the habitat allow each crew member moments of solitude to decompress, reflect, and recalibrate.

To help manage mental well-being, the crew has access to virtual reality tools that simulate Earth environments. Whether it's a walk through a forest, a beach sunset, or even a virtual conversation with family, these experiences provide a comforting escape from the stark reality of Mars. Weekly check-ins with psychologists on Earth help the crew process their feelings, work through conflicts, and prevent potential mental health issues from escalating. Techniques such as journaling, meditation, and structured group discussions are encouraged to maintain open communication and foster a supportive environment.

In a setting where adversity is the norm, psychological resilience is the bedrock of survival. The crew learns to find strength in their mission's purpose, using shared goals and mutual reliance to overcome moments of doubt or fear. The emotional bonds formed through these experiences create a sense of family among the team, offering each member a source of strength when times are tough. By facing adversity together, they reinforce not only their own resilience but also the foundation of a community built to withstand the challenges of an alien world.

These protocols and strategies for handling both physical and psychological crises are essential for life on Mars. Each medical emergency, each emotional struggle, and each moment of adversity transforms the crew, deepening their resolve and uniting them in their commitment to thrive in one of the most challenging environments known to humanity. Through these experiences, the crew embodies the strength, adaptability, and resilience required to push humanity's boundaries, proving that even on the most distant frontier, the human spirit remains unbreakable.

Chapter 12: Looking Forward – Lessons Learned in the First Year

As the first year on Mars draws to a close, the colony stands as a testament to human adaptability and ingenuity. Each challenge faced, each lesson learned, and every system tested has laid a foundation for transitioning into the second year with confidence and resilience. The first year was all about survival—establishing a foothold, proving that humans could endure the Martian environment, and discovering what adjustments were needed for the realities of life on an alien world. Now, as the crew looks ahead, the focus shifts from short-term survival to long-term settlement, evolving the colony from a temporary outpost into a sustainable community.

The experiences of the first year have revealed the crucial adaptations and innovations needed to thrive on Mars. Technologies that have proven most useful, such as the scaled-up MOXIE oxygen generator, Martian concrete for habitat expansion, and advanced water recycling systems, become the pillars of the colony's infrastructure. The MOXIE units, once small-scale experimental devices, have grown into a reliable source of breathable air, reducing the need for oxygen shipments from Earth. This continuous oxygen production now allows the colony to support not only its current residents but also any future arrivals, moving a step closer to complete independence from Earth's resources.

In-situ resource utilization (ISRU) has also demonstrated its power, enabling the crew to manufacture essential building materials from Martian soil. The development of Martian concrete has allowed for rapid construction of radiation-shielded habitats and storage facilities, expanding the colony's living and working spaces. With 3D printing techniques fine-tuned for Martian materials, the crew can now construct everything from structural beams to tool parts, conserving Earth-supplied resources and fostering a new level of self-sufficiency. This building capability will be instrumental in supporting a growing population, as new settlers join the colony over the coming years.

Another key innovation has been the robust water recycling and purification system. Over the course of the first year, the crew has refined this technology to the point where nearly every drop is reclaimed and repurposed. From waste water to condensed atmospheric moisture, this closed-loop system has proven invaluable, supporting both the crew's daily needs and future agricultural efforts. With the colony increasingly capable of generating its own water from Martian ice, reliance on Earth is minimized, bringing Mars closer to becoming a truly self-sustaining environment.

The technological advancements and methodologies developed during this inaugural year also extend to energy management. The hybrid power system, combining solar arrays with experimental compact nuclear reactors, has been rigorously tested under Mars' harsh conditions. The ability to store and manage energy effectively during dust storms and long nights has provided the colony with stability, ensuring that essential life-support and operational systems remain uninterrupted. This energy independence is a cornerstone for future expansion, supporting everything from extended rover missions to advanced greenhouse facilities that will deepen the colony's agricultural capabilities.

Beyond technology, the crew's social dynamics and psychological resilience have also evolved, adapting to the unique pressures of Martian life. They have become adept at navigating isolation, managing limited personal space, and sustaining mental health through routine, shared purpose, and innovative tools like virtual reality environments. These social adaptations are invaluable lessons for the incoming settlers, as the original crew's experiences will serve as a blueprint for how future Martians can live, work, and thrive together.

Each of these innovations—whether in oxygen production, water management, habitat construction, or psychological support—reflects the adaptability and endurance of humanity. As they transition into year two, the crew isn't just surviving on Mars; they are setting the stage for a long-term human presence, proving that a sustainable life on Mars is achievable. These foundational efforts provide the groundwork for generations of explorers, scientists, and settlers who will follow, gradually transforming Mars from an outpost into a home, a place where humanity's frontier spirit continues to reach new heights.

With the success of the first year, Mars is no longer just a distant dream—it's a tangible frontier where humanity has begun to take root. As the colony transitions into its second year, attention turns to what lies ahead: the arrival of subsequent waves of colonists, each bringing new skills, knowledge, and resources to expand and fortify this growing Martian society. These future missions are meticulously planned, with each new group selected to fulfill specific roles that build upon the accomplishments of the original crew, supporting the colony's evolution from a fragile settlement into a self-sustaining community.

Subsequent waves of colonists will include engineers, botanists, medical specialists, and scientists, each contributing to the layered goals of survival, research, and sustainability. Engineers will focus on scaling the habitat infrastructure, constructing larger living quarters, and implementing advanced energy systems to accommodate an expanding population. Botanists and agricultural experts will work to diversify and increase crop yields, aiming to reduce reliance on Earth for food. Medical specialists will enhance the colony's health infrastructure, providing advanced care capabilities and studying the effects of long-term life on Mars to better prepare future settlers. Scientists will push forward with geological studies, atmosphere testing, and potential discoveries that might change our understanding of Mars itself.

In parallel, these new arrivals will bring fresh perspectives and ideas that drive innovation, pushing the boundaries of Martian technology. Each group will contribute to the ongoing adaptation of systems initially developed by the first settlers, refining water reclamation, waste management, and resource extraction techniques to accommodate the demands of a growing colony. This expanded infrastructure is pivotal to the colony's evolution, setting the stage for a Mars that is no longer just an outpost but a thriving settlement capable of supporting a diverse and dynamic population.

The long-term vision for Mars extends beyond merely surviving; it is about thriving, expanding, and establishing a human presence that could someday rival our existence on Earth. Mars represents a new chapter in human exploration, where each success on the red planet fuels ambitions to go even further. The lessons learned and technologies developed on Mars lay the groundwork for humanity's future on other planets, moons, and even deep space. By learning to live off the land on Mars, harnessing local resources, and creating systems for autonomy, humanity builds the tools and mindset necessary to explore the solar system—and eventually, the stars.

The colony on Mars also serves as a model for international cooperation and scientific progress, where the common goal of exploring beyond Earth unites countries, industries, and cultures. Future generations may look back on these early missions as the foundation of an interplanetary society, where knowledge, innovation, and resilience paved the way for a multi-planetary human race. The settlers on Mars carry the legacy of human exploration, inspiring future missions that will bring humanity closer to a universe where life on multiple worlds is not only possible but inevitable.

As Mars transforms from an isolated outpost to a thriving colony, it redefines what humanity can achieve, inspiring us to view the cosmos as our potential home. This long-term vision is more than just a distant goal; it's a reminder of the boundless possibilities that arise when human curiosity and determination reach beyond our own world. Each step taken on Mars opens the door wider, illuminating a path forward, from our own solar system to the uncharted realms beyond, where the spirit of exploration promises to carry humanity to the next frontier.

Conclusion

The first year on Mars marks a monumental achievement, not only for those who lived through its challenges but for all of humanity. In enduring the harsh conditions, mastering the basics of survival, and establishing the first foothold on another planet, the Martian settlers have redefined what is possible. Their journey is more than just a technical feat; it represents the unyielding human spirit that seeks to explore, learn, and expand its horizons. Surviving and even thriving through the initial year is a testament to the courage, ingenuity, and determination that brought them to Mars, echoing the spirit of explorers from generations past who ventured into the unknown with little more than hope and resilience.

This endeavor offers profound lessons for life on Earth, reshaping our understanding of resourcefulness and resilience. On Mars, every drop of water is precious, every watt of power accounted for, and every material maximized to its fullest potential. This mindset of conservation and efficiency, born out of necessity on Mars, offers a template for a more sustainable future back home. As Earth grapples with its own challenges, the Mars experience reminds us of the value in rethinking resource management, adapting to changing environments, and embracing innovative solutions that can sustain life. The skills honed in the pursuit of Mars colonization—adaptability, patience, and collaboration—are as relevant on Earth as they are millions of miles away, highlighting the interconnectedness of human survival across any frontier.

More than a destination, Mars now stands as a launchpad for future exploration. With its rich resources and position within the solar system, Mars offers a stepping stone for missions that will venture deeper into space. The skills, technologies, and knowledge gained from surviving on Mars will fuel missions to the outer planets and beyond, where the challenges are greater and the unknowns even more vast. By establishing a presence on Mars, humanity has proven its ability to adapt to extraterrestrial environments, setting the stage for an era of exploration that reaches well beyond our nearest planetary neighbor. Each success on Mars serves as a foundation for the bold, ambitious journeys yet to come.

As humanity looks to the journey ahead, the first year on Mars becomes a beacon of our interplanetary aspirations. It symbolizes the start of a new chapter, where exploration and survival on other worlds are no longer the realm of science fiction but a tangible reality. Mars is just the beginning—a place where we learned to live beyond Earth, to adapt, and to grow. Each future mission, each new outpost, and each discovery will build upon this foundation, carrying humanity further into the cosmos.

In surviving and thriving on Mars, humanity has taken its first true steps toward becoming an interplanetary species. The Martian settlers have proven that we can not only reach new worlds but also make them our own. The journey is far from over; in fact, it has only just begun. But as we look forward, inspired by the challenges conquered and the horizons that await, the path to the stars seems a little closer, our ambitions a little more achievable. Humanity's new frontier beckons, and with each step, we realize that our journey through the universe is only limited by our courage and imagination.

www.ingramcontent.com/pod-product-compliance
Lightning Source LLC
Chambersburg PA
CBHW070251220526
45465CB00004B/1572